DATE DUE		
JAN 11 '64		
JAN 8 '66		
MAY 10 '69		
DEC 8 '71		
MAY 1 '70		
GAYLORD		PRINTED IN U.S.A.

COAL TRAFFIC ON THE OHIO RIVER SYSTEM

by
FRED A. CARLSON, *Professor*
and
FRANK SEAWALL, *Associate Professor*
Department of Geography
The Ohio State University

Bureau of Business Research Monograph Number 107

Published by
BUREAU OF BUSINESS RESEARCH
College of Commerce and Administration
THE OHIO STATE UNIVERSITY
COLUMBUS, OHIO

COLLEGE OF COMMERCE AND ADMINISTRATION
James R. McCoy, *Dean*

Bureau of Business Research Staff
Viva Boothe, *Director*
James C. Yocum, *Associate Director*

Mikhail V. Condoide, *Economics*
Paul G. Craig, *Economics*
J. L. Heskett, *Management*

Ralph M. Stogdill, *Personnel*
Martha Stratton, *Statistics*
Omar Goode, *Tabulations*

Research Assistants

Ronald Brady
Jeanette Filsinger
Mary Martin

James Neuhart
David Potter
Roderick Purcell

Jeanne Davis

Copyright 1962
The Ohio State University

FOREWORD

Man's oldest mode of mass movement of commodities, inland water transportation, remains relatively important in our modern urban-industrial society. At present inland water transport is specialized, both in the area served and the type of commodities transported. The Nation's waterways are limited in length when compared with the other types of transportation. However, the waterways have the greatest traffic density of all carriers when the comparison is based on ton-miles of traffic per mile of route. Certain commodities have physical characteristics which favor water transportation.

An important segment of our major inland waterways is the Ohio River System. Its navigable rivers penetrate some of the Nation's most productive coal deposits, and coal is its single most important item of traffic. The coal is consumed in large quantities not only by some of the industrial facilities located adjacent to the waterways but also is transhipped by rail to more distant markets.

The major purpose of this study is to show the interrelationships of the Ohio River System and coal traffic. Emphasis is placed on the extent of coal reserves in the counties adjacent to the Ohio River and its tributaries, in order to understand the present, and to develop basic concepts for the future coal traffic potential.

VIVA BOOTHE, Director

TABLE OF CONTENTS

Section	Page
I. Water Transportation and Coal	1
The Ohio River System	1
Improvement in Navigation Facilities	6
II. Coal Reserves	9
Reserves and Access to Water Transportation	13
III. Coal Production	21
Northern Appalachian Coal Field	21
Eastern Interior Coal Field	27
Middle Appalachian Coal Field	28
Southern Appalachian Coal Field	29
IV. Government Regulation of River Traffic	31
Rates and Costs of Transportation of Coal	32
V. Traffic Flow	37
Upper Ohio River	38
Middle Ohio River	39
Lower Ohio River	41
VI. Consumption of Bituminous Coal	43
VII. Summary	49

LIST OF TABLES

Table Number   *Page*

1. Bituminous Coal Produced in the United States and in the Ohio Valley States, 1950-1959 2
2. Bituminous Coal Reserves in Ohio Valley States, as of January 1, 1960 .. 11
3. Estimated Coal Reserves in Pennsylvania in Counties Bordering Ohio River Waterways 13
4. Estimated Bituminous Coal Reserves in Ohio in Counties Bordering Ohio River Waterways 14
5. Estimated Coal Reserves in West Virginia in Counties Bordering Ohio River Waterways 16
6. Estimated Reserves in Indiana in Counties Bordering Ohio River Waterways 17
7. Estimated Coal Reserves in Illinois in Counties Bordering Ohio River Waterways 18
8. Estimated Coal Reserves in Eastern Kentucky in Counties Bordering Ohio River Waterways 19
9. Estimated Bituminous Coal Reserves in Tennessee in Counties Bordering Ohio River Waterways 20
10. Bituminous Coal Shipped on the Ohio River System in 1959, by River and County of Origin 24
11. River and Rail Rates per Ton for the Transportation of Coal Between Selected Points 32
12. Ton-Mile Common Carrier Rates for Coal and Distances for River and Rail Between Selected Points 32
13. Major Consumers of Coal in the United States 45
14. Fuel Economy in Consumption of Coal at Electric-Utility Power Plants 46

LIST OF FIGURES

Figure Number — *Page*

1. Bituminous Coal Mined in the United States, 1950-1959 3
2. Coal Traffic in Relation to Total Traffic on the Ohio River System, 1959 4
3. Face Lifting on the Ohio .... Opposite Page 6
4. Bituminous Coal and Inland Waterways 10
5. Bituminous Coal Reserves in Ohio Valley States January 1, 1960 12
6. Shipment of Bituminous Coal, by Counties Bordering Ohio River Waterways, 1959 26
7. Relative Changes in Bituminous Coal Traffic on Ohio Waterways and U. S. Coal Production, 1950-1959 33
8. Coal Traffic Density on the Ohio River System, 1959 ... Opposite Page 38
9. Origin of Coal Shipments on Ohio River System, 1950-1959 39
10. Percentage of Total Consumption of Bituminous Coal, by Type of Consumer in the United States, 1950-1959 43
11. Consumption of Coal at Electrical-Utility Power Plants in the United States, 1920-1959 44

I

WATER TRANSPORTATION AND COAL

The industrial growth along the Ohio River and its tributaries during the years 1950–1959 represented a capital expenditure of $15 billion. This impressive development is attributed, partly, to the availability of coal and low-cost river transportation. In 1959, the production of bituminous coal in states bordering the navigable waterways of the Ohio River System exceeded 361,000,000 tons, or nearly 88 per cent of the Nation's total output of bituminous coal, as illustrated by Figure 1 and Table 1. The coal reserves in the area are tremendous, estimated at 458,600,000,000 tons. In 1959, nearly 42,780,000 tons of coal traffic originated on the 2,700 miles of navigable waterways which serve this rich and productive region, "The Ohio River Valley."[1] In 1959 more than 143,000,000[2] tons of commodities were transported over the waters of the Ohio River System, one of the world's great inland waterways. On the main stem of the System, the Ohio River, a navigable waterway 981 miles long with a minimum channel depth of 9 feet, from Pittsburgh to Cairo, the total volume of traffic increased from 42,792,000 tons in 1948 to 80,800,000 tons in 1959, with an all time high of 82,000,000 tons in 1957. Coal, petroleum and products, sand and gravel, iron and steel, and industrial chemicals are the leading commodities transported on the Ohio River System.

THE OHIO RIVER SYSTEM

Although the Ohio River ranks high in its commercial importance, its greatness, however, must be considered in relation to its major "Contributories," the Tennessee, the Cumberland, the Green, the Kentucky, the Kanawha, the Monongahela, and the Allegheny. The total mileage of these tributaries with a channel depth of 9

---

[1] The Ohio River Valley states included in this report are the states which are adjacent to the Ohio River and its navigable tributaries. These states include: Alabama, Illinois, Indiana, Kentucky, Ohio, Pennsylvania, Tennessee, and West Virginia.

[2] A portion of this traffic is shipped over more than one river, hence there is some duplication in this figure.

TABLE 1—Bituminous Coal Produced in the United States and in the Ohio Valley States, 1950–1959
(In Thousands of Tons)

State	1950	1951	1952	1953	1954	1955	1956	1957	1958	1959
Alabama	14,422	13,597	11,383	12,532	10,282	13,088	12,663	13,260	11,182	11,947
Illinois	56,291	54,200	45,790	46,010	41,971	45,932	48,102	46,993	43,912	45,466
Indiana	19,957	19,451	16,350	15,812	13,400	16,149	17,089	15,841	15,022	14,804
Kentucky	78,495	74,972	66,144	65,060	56,964	69,020	74,555	74,667	66,312	62,810
Ohio	37,761	37,949	36,209	34,737	32,469	37,870	38,934	36,862	32,028	35,112
Pennsylvania	105,870	108,164	89,181	93,331	72,010	85,713	90,287	85,365	67,771	65,347
Tennessee	5,070	5,401	5,265	5,467	6,429	7,053	8,848	7,955	6,785	5,913
West Virginia	144,116	163,310	141,713	134,105	115,996	139,168	155,891	156,842	119,468	119,692
Total	461,982	477,044	412,035	407,054	349,521	413,993	446,369	437,785	362,480	361,091
Other States[a]	54,329	56,621	54,806	50,236	42,185	50,640	54,505	54,919	47,966	50,937
U. S. Total[a]	516,311	533,665	466,841	457,290	391,706	464,633	500,874	492,704	410,446	412,028

[a] Includes lignite.
Source: Bureau of Mines.

feet or more is greater than 1,000 miles. Based on total volume, the tributaries are relatively important; in 1959 the total waterborne traffic on the tributaries exceeded 62,200,000 tons compared to 80,800,000 tons on the Ohio River. Approximately 42 per cent of the total freight traffic on the Ohio River originates on the tributaries

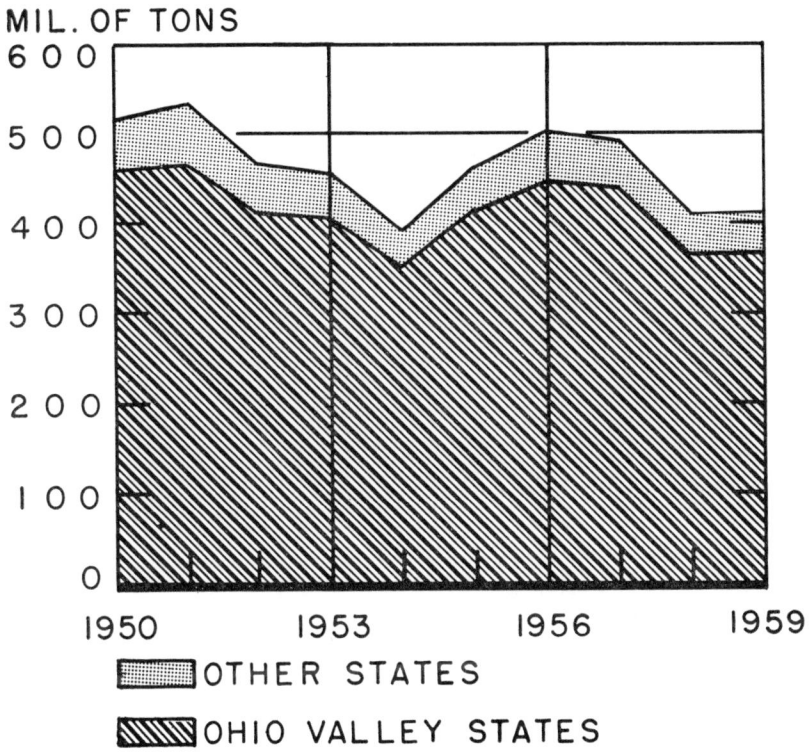

FIGURE 1—Bituminous Coal Mined in the United States, 1950–1959

Source: Bureau of Mines.

or the Mississippi River, and about 57.2 per cent of this traffic is coal. In fact, the total coal traffic on the Ohio River in 1959 was slightly less than the coal traffic on the tributaries, 40,755,000 tons on the Ohio River compared to 40,982,000 tons on the tributaries (Figure 2).

FIGURE 2—Coal Traffic in Relation to Total Traffic on the Ohio River System, 1959

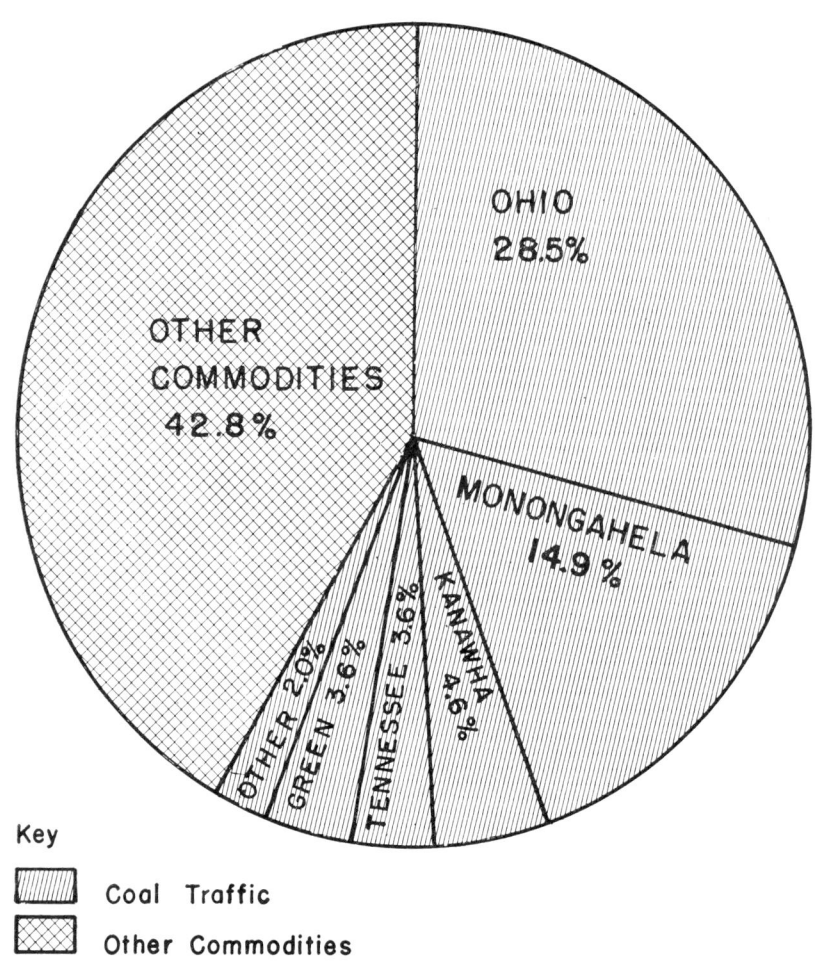

Traffic on each of the tributaries of the Ohio River can be identified by one or more characteristics. The Tennessee River is one of the great tributaries of the Ohio River. Its commercial position has been somewhat overshadowed by the Tennessee Valley Authority development of power plants and flood control dams. Over its course of some 652 miles with a minimum channel depth of 9 feet from Knoxville, Tennessee, in the mountains to its confluence with the Ohio River at Paducah, Kentucky, the barge traffic in 1959 exceeded 12,000,000 tons, including coal, grain, petroleum products, steel, and industrial chemicals. The coal traffic was over 5,000,000 tons of which 3,000,000 tons moved upstream to TVA steam-electric plants. The Cumberland River with its limited channel depth transports little or no coal. Its cargo consists, primarily, of motor fuel and other petroleum products, metals, manufactures, and industrial chemicals.

With respect to total traffic and particularly coal traffic the Monongahela is by far the most important tributary of the Ohio River. More coal is transported on the Monongahela River than on all the other tributaries combined. It is likely that the coal traffic density on the Monongahela River is greater than that on any other river in the world. In 1959, over 21,000,000 tons of coal were shipped on this river which is formed by the confluence of the West Fork and Tygart Rivers, and is navigable for its entire length, 129 miles. Most of this coal, which is of coking quality, is shipped downstream, destined for the by-product coke ovens in the steel plants located adjacent to the lower Monongahela and upper Ohio Rivers. There are also downstream shipments of steam coal for steam-electric plants. Coal comprises slightly over 78 per cent of the total Monongahela River traffic. Other commodities in the traffic flow are iron and steel, sand and gravel, fuel oil and gasoline, and chemicals.

The Allegheny River has had a long history as a transporter of coal from mines near the lower river to downstream markets on the upper Ohio River. In 1959, coal shipments were nearly 2,000,000 tons, which constituted nearly 50 per cent of the river traffic. Other major commodities included in the Allegheny River traffic are sand and gravel, iron and steel, gasoline, and petroleum asphalt.

By comparison with the other Ohio River tributaries coal traffic on the Kanawha River is second only to the Monongahela River. In 1959, coal shipments totaled 6,600,000 tons which comprised approximately 68 per cent of the Kanawha River traffic. The basic traffic pattern for coal is downstream, destined for Ohio River markets.

Other major commodities transported on the 90-mile navigable segment of the Kanawha River include chemicals, fuel oil and gasoline, and sand and gravel.

Over 99 per cent of the traffic on the Green River is coal. In 1959, coal shipments were in excess of 5,100,000 tons. Most of this traffic was downstream, destined for steam-electric plants on the shores of the lower Ohio River.

Small coal shipments also originate on the Kentucky, Little Kanawha, and Tradewater Rivers. But by comparison, with the other tributaries in 1959, the traffic was relatively limited. Each of these rivers transported over 100,000 tons of coal, but none exceeded 150,000 tons.

IMPROVEMENTS IN NAVIGATION FACILITIES

Significant improvements in navigation facilities are in progress on the Ohio River System. The master plan calls for a reduction of the number of lock and dam installations from the original 46 to 19. The new structures at each site will consist of a gate-controlled dam and dual locks; one lock, 110 feet by 1,200 feet, and the other, 110 feet by 600 feet. The latter, the auxiliary lock, is similar to the existing locks. The new developments are the New Cumberland (recently completed) and Pike Island located within 90 miles of Pittsburgh; the Greenup in the Huntington area; the Anthony Meldahl and the Markland near Cincinnati; and the McAlpine, currently known as Lock 41 at Louisville. It is reported that by 1965, all 6 of these new locks and dams will be completed (Figure 3).

By increasing the length of the pools through the reduction in the number of dams, considerable time and related costs will be saved by the barge operators. For example, the Markland project will replace 5 old installations with a pool length of about 95 miles as compared to the present average pool of less than 20 miles during low water season. For a tow with a speed of 7.5 miles per hour, the time to navigate this section will be reduced from 20 hours to 13 hours, allowing for passage through the Markland Lock. It has been estimated that the cost of operating a representative tow the full length of the Ohio River will be reduced by some $1,500 when all 6 of the new installations have been completed.

FIGURE 3—Face Lifting on the Ohio

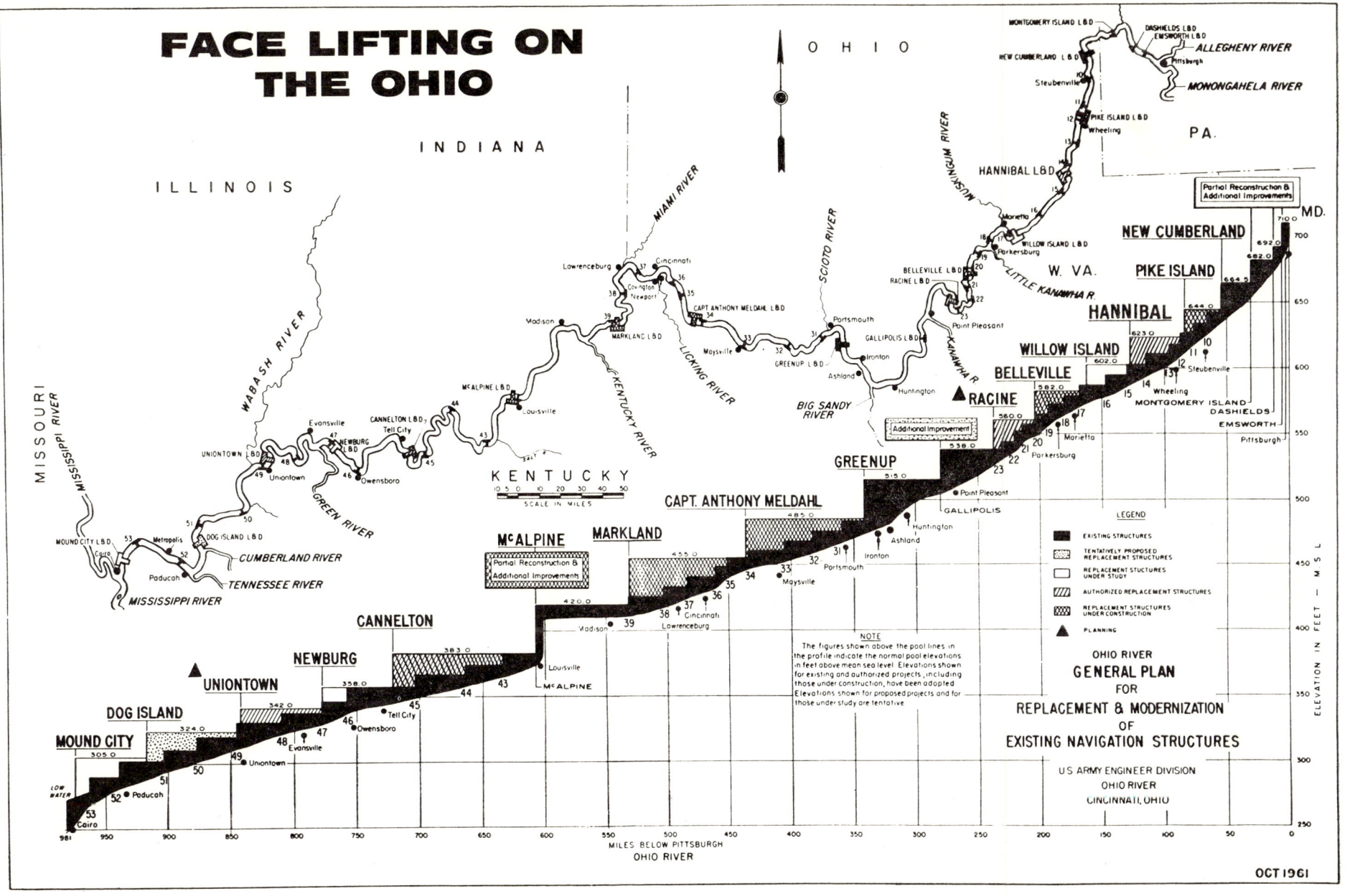

Waterway improvements are also in process on some of the Ohio River tributaries which will be a distinct benefit to transportation. When the Opekiska Lock and Dam on the upper Monongahela River is completed, a 9-foot navigable channel will be maintained to the headwaters of the Monongahela River. The Opekiska project will replace two old locks that have lock chambers which measure but 56 by 182 feet and have a depth of 7 feet over the lock sills. The completion of the new Maxwell Lock and Dam, also on the Monongahela River, will increase the size of the lock chamber and will facilitate the movement of traffic through the busy middle segment of the river. Navigation on the Cumberland River will be improved with the completion of the new multipurpose Barkley Lock and Dam.

Other projects under consideration include the development of the Wabash River of Indiana and the Big Sandy River of eastern Kentucky with a series of locks and dams.

Along with the modernization of the navigational properties of the Ohio River System emphasis is in evidence on the improvement of the equipment, such as the increase in the power of the towboats and the size of the barges, and construction of special purpose carriers. Fifteen years ago a towboat of 3,200 horsepower represented the maximum power. Now towboats with horsepower of 4,800 are common. The newest diesels deliver 6,000 or more horsepower and can push 20 to 40 barges loaded with 15,000 to 35,000 tons of cargo.

Barges have been increased in size and carrying capacity. The early coal barge, the so-called "Pittsburgh Standard," an open-hopper type measured 175 feet by 26 feet, had a maximum capacity of about 1,000 tons. It was followed by the "Jumbo Barge," with dimensions 195 feet by 35 feet and a maximum capacity of 1,500 tons. A larger barge, 290 feet by 50 feet with a capacity of 3,200 to 3,500 tons, has been designed, primarily for long-distance hauls. Special purpose barges, also, are in use, designed to haul certain chemicals, motor vehicles, and petroleum products.

The increased power of the towboats, the length of the pools, the size of the lock chambers, the size of the barges, and the length of the tows have improved the water transportation on the Ohio River System.

II

COAL RESERVES

The Ohio River and its navigable tributaries provide ready access to large reserves of quality coal. Due to the location of the Ohio River System, as illustrated by Figure 4, water transportation is provided for many areas of the Appalachian Coal Province and parts of the Eastern Interior Coal Field in the Interior Province. Coal shipments originate in several segments of the Ohio River and all of its major navigable tributaries[1], except the Cumberland River.

Although the largest coal reserves are in the western part of the United States, the coal reserves of the Appalachian and Eastern Interior Coal Fields are of superior quality and are of sufficient size to supply the regional market for many years. The estimated original bituminus coal reserves in the states adjacent to the Ohio River waterways exceeded 500,000,000,000 tons as illustrated by Table 2 and Figure 5. As of January 1, 1960, these states had coal reserves of approximately 458,600,000,000 tons. Not all of the reserves can be classified as recoverable coal; several factors must be considered before recoverable coal reserves can be estimated. First, in underground mining some of the coals are undisturbed to prevent collapse of the mine roof. Some coal beds included in the original estimates of coal reserves may be too thin to make mining economically feasible. In the classification of coal reserves relative to the thickness of the seam, coal seams which are less than 14 inches thick are not included. The presence of excessive parting material within the coal seam, such as shale or slate may prohibit mining of some coals as cleaning costs may be excessive. Unusual depth of the coal seam may prohibit mining because of high cost to get a shaft to the coal bed, and excessive ventilation costs. Faulting and folding of coal seams may make mining extremely difficult and unprofitable. The presence of large amounts of water which must be pumped out, or excessive amounts of gas which may increase ventilation costs and

---

[1] Major navigable tributaries include: the Monongahela, Kanawha, Green, Tennessee, Allegheny, Kentucky, and Cumberland Rivers. Navigable segments of these rivers are at least 70 miles in length. Coal shipments also originate on some of the smaller navigable tributaries which include the Little Kanawha, Tradewater, Clinch, and Emory Rivers.

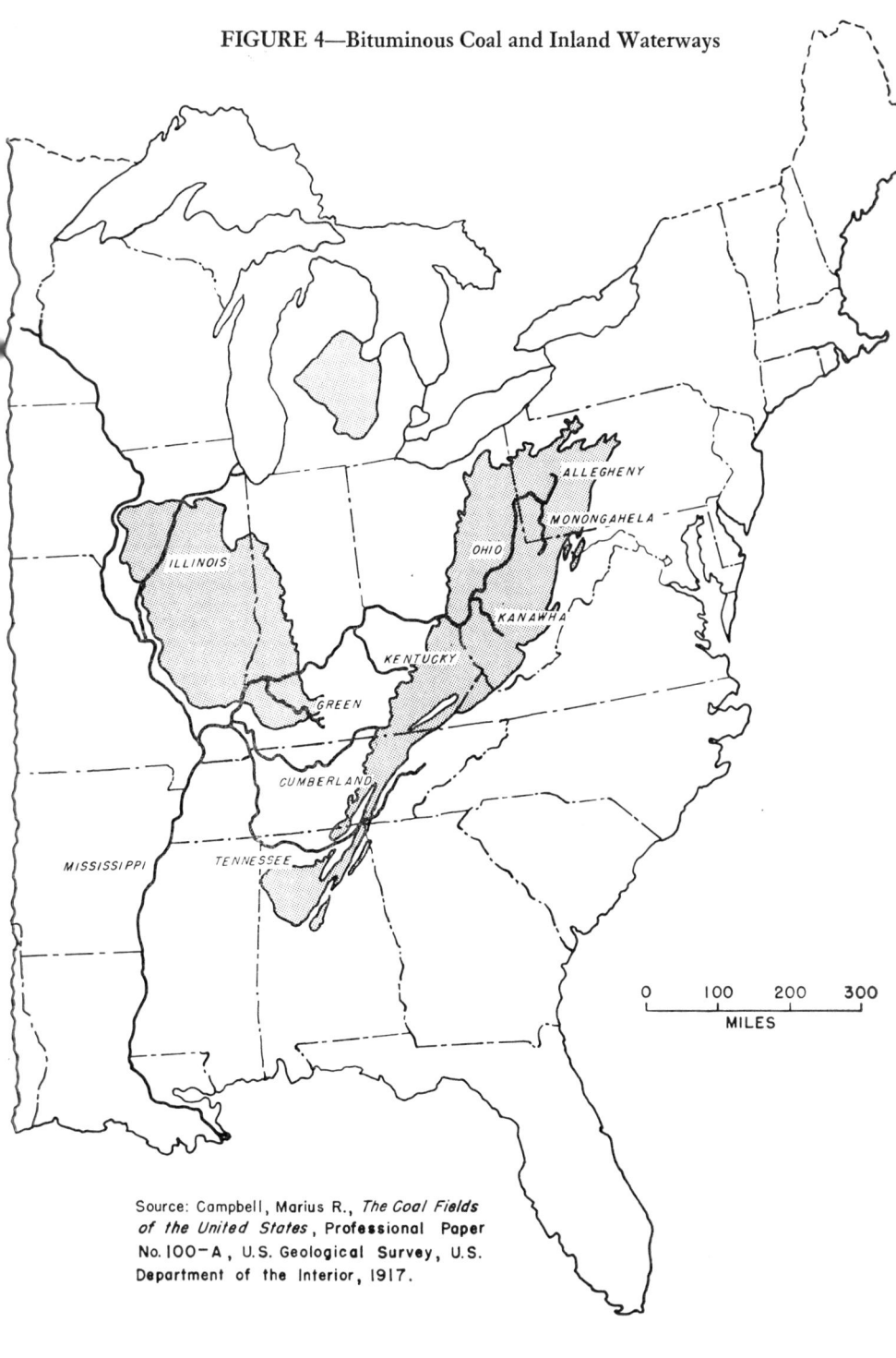

FIGURE 4—Bituminous Coal and Inland Waterways

Source: Campbell, Marius R., *The Coal Fields of the United States*, Professional Paper No. 100-A, U.S. Geological Survey, U.S. Department of the Interior, 1917.

limit the use of explosives in mining operations may prohibit the mining of some coal beds. Due to these variable factors the United States Bureau of Mines assumes that about one-half of the coal reserves can be mined at the present level of technology. Therefore, in the states bordering the Ohio River Waterways, 92 per cent of the original coal reserves, or 458,600,000,000 tons remain in this region, of which one-half is considered recoverable.[2] Using the Bureau of

TABLE 2—Bituminous Coal Reserves in Ohio Valley States, as of January 1, 1960
(In Millions of Short Tons)

State	Estimated Original Reserves	Estimated Production Plus Loss in Mining	Remaining Reserves	Recoverable Reserves Jan. 1, 1960, Assuming 50 Per Cent Recovery
Alabama	13,754[a]	46[b]	13,708	6,854
Illinois	137,329[c]	948[d]	136,381	6,190
Indiana	37,293	2,296	34,997	17,499
Kentucky	72,318	5,292	67,026	33,513
Ohio	46,488	4,104	42,384	21,192
Pennsylvania	75,093	16,736[e]	58,357[e]	29,178[e]
Tennessee	1,912[f]	12[g]	1,900	950
West Virginia	116,618	12,738	103,880	51,940
Total	500,805	42,172	458,633	229,316

[a] Remaining reserves, Jan. 1, 1958.
[b] For 1958 and 1959 only.
[c] Remaining reserves, Jan. 1, 1950.
[d] For 1950 through 1959.
[e] Estimated by authors.
[f] Remaining reserves, Jan. 1, 1959.
[g] For 1959 only.
Source: Bureau of Mines.

Mines' standards, approximately 229,300,000,000 tons of coal are estimated as recoverable as of January 1, 1960. These coal deposits can be considered as one of the major reserves of energy in the world.

Strip mines normally report a coal recovery rate in excess of the 50 per cent associated with underground mines. In strip mines, the usual recovery rate is 80 per cent.[3] Strip mines have a higher rate of recovery since it is not necessary to leave undisturbed pillars of coal to support the roof as in underground mining.

Since all coal reserve data are estimates, an explanation of the methods used in the classification of data is necessary. The Bureau of

---

[2] With improved mining technology, the 50 per cent rate appears to be a conservative estimate.
[3] Spencer, Frank D., *Coal Resources of Indiana*, Geological Survey Circular 266, (Washington: Government Printing Office, 1953), p. 39.

Mines' classification includes three categories of coal reserves; measured, indicated and inferred. The measured reserves are based on observations of outcrops, mine workings and drill holes. Assays of this type are usually quite accurate as errors are normally less than

FIGURE 5—Bituminous Coal Reserves in Ohio Valley States, January 1, 1960

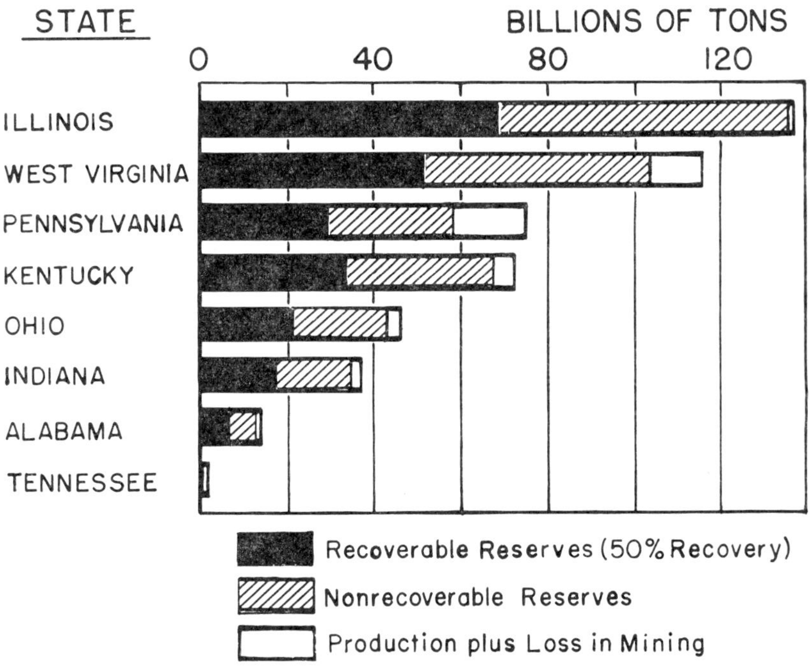

- ■ Recoverable Reserves (50% Recovery)
- ▨ Nonrecoverable Reserves
- ☐ Production plus Loss in Mining

Source: Table 2.

20 per cent. Indicated reserves are computed partly on direct measurements and partly from a projection of visible data based on geologic evidence. Accuracy of estimates for indicated reserves are slightly less than for the measured reserves. Inferred reserves are based on widely spaced drill records and assumptions of continuity of beds. These are the least accurate. In this study, no attempt has been made in an analysis of the classification of the types of coal reserves since this information is not available for all areas.

RESERVES AND ACCESS TO WATER TRANSPORTATION

In order to show more specifically the coal reserves with access to water transportation in the Ohio Valley, tables have been prepared, based on county data. In some coal producing areas, namely western Kentucky and northern Alabama, coal reserve data by counties were not available. With the present development of supplemental transport such as conveyor belts, mine cars, and trucks, it is logical to assume that most of the coal in a county bordering a navigable river will have access to water transportation.

As shown in Table 3, Greene, Washington, Fayette, and Westmoreland counties in Pennsylvania have significant coal reserves. However, these data are based on a 1944 estimate and there has

TABLE 3—Estimated Coal Reserves in Pennsylvania in Counties Bordering Ohio River Waterways
(In Thousands of Tons)

County	Original Reserves	Mined or Lost In Mining	Estimated Remaining Reserves As of January 1, 1945
Allegheny	3,515,000	1,370,000	2,145,000
Armstrong	4,124,000	220,000	3,905,000
Beaver	1,390,000	17,000	1,373,000
Fayette	7,565,000	2,000,000	5,565,000
Greene	10,610,000	173,000	10,437,000
Washington	11,580,000	1,020,000	10,560,000
Westmoreland	7,020,000	1,595,000	5,425,000
Total	45,805,000	6,395,000	39,410,000

Source: Ashley, George H., et al., *Pennsylvania's Mineral Heritage*, Pa. Topographic and Geological Survey, 1944, p. 83.

been a considerable amount of coal produced from these counties since this assay was completed. Drift mining has depleted some of the coal near the Monongahela River as these areas were the first to be mined since they had ready access to water transportation. To date, most of the production has been from the Pittsburgh Coal Seam, which is of coking quality, but this area has sizable deposits of other coal beds such as: the Sewickley, Redstone, Upper and Lower Kittanning, Upper and Lower Freeport, and several other beds. Most of these seams are of lower quality than the famous

Pittsburgh Coal Seam, but these coals are classified as bituminous and can be used for many industrial purposes. At present, Greene County in the southwest corner of the state has some excellent deposits of quality coal with access to Monongahela River transportation. The Nation's largest coal mine, the Robena Mine, is located in the eastern part of Greene County and utilizes Monongahela River transportation to ship its coal to market.

Although there is some strip mining in southwestern Pennsylvania most of the coal is recovered by drift and shaft mining. The structure is such that most of the coal beds are nearly horizontal with very little faulting.

Several southeastern Ohio counties, which have access to Ohio River transportation, have sizable coal reserves (Table 4). The esti-

TABLE 4—Estimated Bituminous Coal Reserves in Ohio in Counties Bordering Ohio River Waterways
(In Thousands of Tons)

County	Original Reserves	Production 1800–1958	Recoverable[a] Reserve
Athens	2,225,354	197,280	915,397
Belmont	5,759,456	441,905	2,437,823
Columbiana	2,803,343	58,064	1,343,617
Gallia	1,642,616	9,784	811,523
Jefferson	3,433,759	271,035	1,445,844
Lawrence	1,862,713	14,652	916,704
Meigs	942,190	48,272	422,823
Monroe	2,985,416	138	1,492,570
Scioto	87,252	317	43,309
Washington	1,262,721	2,232	629,128
Total	**23,004,820**	**1,043,679**	**10,458,738**
Estimated Loss in Mining		1,043,665	

[a] Based on 50 per cent Recovery.

Source: Brant, Russell A., and DeLong, Richard M., *Coal Reserves in Ohio*, Ohio Geological Survey, 1960, p. 7.

mated original coal reserves of the counties bordering navigable waterways was over 23,000,000,000 tons from which 1,044,000,000 tons have been produced as of January 1, 1959. Data contained in an excellent study on Ohio coals by Brant and DeLong, entitled *Coal Resources of Ohio,* published by the Ohio Geological Survey in 1960, specifies that approximately one-half of the total coal reserves of Ohio are in counties adjacent to the Ohio River. The Lower and Middle Kittanning and Pittsburgh Coal Seams comprise approximately 54 per cent of Ohio's original coals. Within Ohio, the Pittsburgh Coal Seam is largely concentrated in the counties bordering the Ohio River; this seam, however, diminishes in thickness in its western margin. This trend is characteristic of almost all of the 23 Ohio coal beds. Most of Ohio coals are of moderate thickness; according to the Bureau of Mines, only 4 per cent of the seams measure over 6 feet, and 53 per cent are from 4 to 6 feet thick[4]. Many of the coal beds of Ohio are nearly horizontal and many seams are sufficiently near the surface to permit strip mining. Over 51.5 per cent of the coal production in the counties bordering the Ohio River was produced by strip mining.[5] The recoverable coal reserves of Belmont, Monroe, Jefferson, and Columbiana Counties are the most important of the counties bordering the Ohio River. Belmont County recoverable coal reserves are estimated at over 2,400,000,000 tons, whereas each of the other counties listed above has estimated recoverable coal reserves in excess of 1,300,000,000 tons.

Due to the long river frontage on the Ohio River and penetrations by the Kanawha and Monongahela Rivers, many West Virginia counties have access to water transportation. The estimated original coal reserves of mineable thickness in the 15 counties of West Virginia, with coal deposits along the navigable waterways, totaled over 31,000,000,000 tons (Table 5). As of January 1, 1960, it was estimated that coal reserves totaled over 29,000,000,000 tons.[6]

Coal deposits in northern West Virginia are basically a continuation of the adjacent Pennsylvania coal beds. Monongalia and Marion Counties which are served by the Monongahela River, have relatively large coal reserves, totaling 3,421,000,000 and 3,874,000,000 tons, respectively. It seems logical to assume that the mining of the quality coals in these counties will be expedited with the completion of the two new locks and dams on the upper Monongahela

[4] U. S. Bureau of Mines, *Minerals Yearbook 1956*, Vol. 2, Fuels, (Washington, D.C., 1958), p. 31.
[5] Department of Industrial Relations, State of Ohio, *Annual Coal and Nonmetallic Minerals Report for 1958*, (Columbus, 1958), p. 30.
[6] Department of Mines, State of West Virginia, *Annual Report, 1959*, (Charleston, 1960), p. 108.

River. The 7-foot channel in the upper 15 miles of the Monongahela River will become a 9-foot channel upon completion of this project. Coal reserves in Marshall and Wetzel Counties are in excess of 7,700,000,000 tons. Although there has been some coal mined in each of these counties the total production is insignificant in comparison with the total reserves. Marshall and Wetzel Counties are served by the Ohio River.

The navigable Kanawha River penetrates into Kanawha and Fayette Counties which have sizable coal reserves. As of January 1, 1960, Kanawha County had over 5,400,000,000 tons of coal remain-

TABLE 5—Estimated Coal Reserves in West Virginia in Counties Bordering Ohio River Waterways
(In Thousands of Tons)

County	Estimated Original Coal of Minable Thickness	Production Removed To Jan. 1, 1960	Estimated Remaining Coal of Minable Thickness
Brooke	360,000	76,262	283,738
Cabell	44,167	....	44,167
Fayette	4,420,505	633,674	3,786,831
Hancock	500,000	3,224	496,776
Kanawha	5,814,356	409,725	5,404,631
Marion	4,317,089	443,224	3,873,865
Marshall	4,448,857	57,733	4,391,124
Mason	339,976	11,454	328,522
Monongalia	3,748,631	327,494	3,421,137
Ohio	910,000	78,438	831,562
Pleasants	....	82	....
Putnam	433,090	21,345	411,745
Tyler	948,133	0	948,133
Wayne	1,471,496	6,845	1,464,651
Wetzel	3,321,923	93	3,321,830
Total	**31,078,223**	**2,069,593**	**29,008,712**

Source: *Annual Report*, Department of Mines, West Virginia, 1959, p. 108.

ing and Fayette County coal reserves totaled over 3,700,000,000 tons. In the past these counties have produced over 1,043,000,000 tons of coal.

The Eastern Interior Coal Field occupies parts of three states, Indiana, Illinois, and Kentucky. Generally, these coals are of lower quality than those of the Appalachian Coal Province. However, the coal of the Eastern Interior Field is classified as superior to the coal of other fields within the Interior Coal Province. Most of the coal from the Eastern Interior Field can be classified as bituminous with the greater part ranked as high volatile C bituminous.[7] Based on present technology this coal is of noncoking quality, but is most satisfactory for the generation of electrical power.

The coal beds of Indiana are on the eastern side of a large structural basin which underlies a part of Illinois, northwestern Kentucky, and the southwestern portion of Indiana. Near the margins of the basin, these coals are close to the surface. Due to the limited over-burden, most of Indiana's coal production is by strip mining.

Within Indiana, 5 counties, bordering the Ohio River, have coal deposits (Table 6). It has been estimated that these counties have approximately 25 per cent of the Indiana coal reserves, or slightly

TABLE 6—Estimated Coal Reserves in Indiana in Counties Bordering Ohio River Waterways
(In Thousands of Tons)

County	Original Reserves	Mined or Lost in Mining to Jan. 1, 1951	Estimated Remaining Reserves as of Jan. 1, 1951
Perry	59,206	0	59,206
Posey	5,393,441	0	5,393,441
Spencer	250,372	3,469	246,903
Vanderburgh	2,185,929	19,006	2,166,923
Warrick	1,685,398	122,207	1,563,191
Total	9,574,346	144,682	9,429,664

Source: Spencer, Frank D., *op. cit.*, p. 5.

over 9,400,000,000 tons. Posey County in the southwest corner of Indiana has the largest coal reserves of any county adjacent to the Ohio River. The remaining reserves are estimated to be over 5,393,000,000 tons. The coal seams in Posey County include Coal III, Coal V, Lower Millersburg, and Upper Millersburg. Some of these coals are good quality steam coal, and seams such as Coal V have an average thickness of 5 feet. However, there has been no coal production in Posey County, since coal in the nearby counties is more readily available. Vanderburgh County, immediately east of Posey County has estimated remaining coal reserves of over

[7] Spencer, Frank D., *op. cit.*, p. 3.

2,160,000,000 tons, and are of the same formations as those of Posey County. There has been limited coal production in Vanderburgh County. At present, Warrick County is the most productive coal producing county in Indiana. Remaining coal reserves as of Jan. 1, 1951 in Warrick County was over 1,563,000,000 tons. Much of this coal is relatively near the surface so that strip mining is used. Other Indiana counties bordering the Ohio River which have deposits of minable coal include Spencer and Perry Counties.

In Illinois, Gallatin and Hardin Counties are the only counties which border the Ohio River and have coal deposits (Table 7). The coal reserves of Gallatin County are impressive, original reserves are

TABLE 7—Estimated Coal Reserves of Illinois in Counties Bordering Ohio River Waterways
(In Thousands of Tons)

County	Original Reserves
Gallatin	3,969,930
Hardin	3,598
Total	3,973,528

Source: Cady, G. H., *Minable Coal Reserves of Illinois,* Ill., Geol Sur. Bull. No. 78, 1952.

estimated as exceeding 3,969,000,000 tons. The characteristics of these coals are similar to the adjacent county in Indiana, Posey County. Hardin County coal reserves are meager.

Kentucky coal deposits which border navigable rivers are located in both the eastern and western parts of the state (Figure 4). The eastern Kentucky coals of the Middle Appalachian Coal Field can be transported via several rivers, namely, the Ohio, Kentucky, Cumberland, and Big Sandy. However, the total coal reserves of the counties bordering these rivers are relatively limited when comparisons are made with counties in other states bordering the navigable waterways. Within eastern Kentucky, Lawrence and Boyd Counties had the largest original coal reserves, estimated at 581,000,000 and 456,000,000 tons, respectively (Table 8). Original reserves of Greenup and Lee Counties are estimated to exceed 200,000,000 tons each.

The western Kentucky coal deposits of the Eastern Interior Coal Field are a continuation of the coal seams of Indiana and Illinois. Strip mining is characteristic of this area since the coal seams are near the surface at the margin of the basin. Specific data on coal reserves of the western Kentucky counties are not available.

Four counties adjacent to the upper Tennessee River in eastern Tennessee are reported to have coal deposits. The reserves of these counties, which are listed in Table 9, are not exceedingly impressive when compared with the coal reserves of West Virginia, Pennsyl-

TABLE 8—Estimated Coal Reserves in Eastern Kentucky
in Counties Bordering Ohio River Waterways
(In Thousands of Tons)

County	Original Reserves
Boyd	456,500
Clinton	6,850
Greenup	288,780
Lawrence	581,470
Lee	200,470
Pulaski	80,050
Wayne	38,800
Total	1,652,920

Source: Huddle, John W., *et al., Coal Resources of Eastern Kentucky,* U.S. Dept. of Interior, Geological Survey, 1958, p. 3.

vania, Ohio and Indiana. It should be noted that Tennessee coal reserves include only coal deposits with a minimum thickness of 28 inches, whereas most other states include all coal reserves which exceed 14 inches. Of the four counties which border the Tennessee River the largest reserves are in Marion and Hamilton Counties.

Specific data on coal reserves by counties for the state of Alabama were not available. However, Jackson, Madison and Marshall Counties in the northeastern part of Alabama which border the Tennessee River are reported to have coal deposits.[8]

Each of the 8 Ohio Valley States have coal deposits in counties bordering the navigable waterways of the Ohio River System (Figure 5). However, there is variety in both quantity and quality of these coals. The largest reserves and best quality coal are in counties

[8] Department of Industrial Relations, State of Alabama, *Annual Statistical Report,* Fiscal Year 1958-1959, (Montgomery, 1960), p. 70.

adjacent to the waterways of West Virginia, Pennsylvania and Ohio. Depletion of the readily accessible quality coal has been characteristic of some Pennsylvania counties. Indiana coal reserves are less than those of the three eastern states, but are still impressive in size. It is unfortunate that coal reserve data by county were not available for western Kentucky and Alabama, but if present pro-

TABLE 9—Estimated Bituminous Coal Reserves in Tennessee in Counties Bordering Ohio River Waterways
(In Thousands of Tons)

County	Coal Reserves[a] as of Jan. 1, 1960
Hamilton	27,354
Marion	47,001
Rhea	15,938
Roane	3,193
Total	93,486

[a] In beds with a minimum thickness of 28 inches.

Source: *Information Circular* No. 10, State of Tennessee, Department of Conservation and Commerce, Division of Geology, Nashville, Tenn., 1960.

duction is an indicator, western Kentucky coal reserves may be substantial.

Although all of the coals are classified as bituminous, the chemical composition of the coals varies considerably. Reserves of high and low volatile bituminous coals are located in the Ohio Valley States. This variety of coals permits varied uses. Generally, the better coals, partially based on BTU content, are located in Pennsylvania and West Virginia.

III

COAL PRODUCTION

Within the Ohio River Valley are parts of the two most productive coal provinces in the Western Hemiphere, the Appalachian and the Interior. The 8 states of the Ohio River Valley normally supply from 80 per cent to 90 per cent of the total coal production of the United States. Coal from these states has provided basic energy for the North American Manufacturing Belt for many years. Although coal production in the United States has declined during the decade of the 1950's (Figure 1), coal is still extremely important to certain types of industry in the Nation.

The Appalachian Coal Province, the world's most productive coal region, is divided into three major fields: the Northern, Middle and Southern. Certain characteristics of this Province are advantageous to coal production. Most of the coal seams are relatively horizontal or have a slight dip. By comparison with other coal mining areas, the coal seams are moderately thick. In some places these coal beds have a thickness in excess of 7 feet. The presence of firm rock overlying the coal seam provides a strong roof which is a distinct advantage for mechanization of coal mining. The Province, one of the most mechanized underground coal producing regions, has one of the highest outputs of coal per miner in the world. The coal deposits in the fields of the Appalachian Coal Province appear in a series of beds, so that if shaft mining is used, several coal seams may be mined from a single shaft. The rolling topography of the Appalachian Plateaus often exposes seams of coal which may be drift mined and loaded directly on river barges located on the navigable streams in the valleys. Although there are chemical differences in the various coals produced in the Appalachians, all of this coal is classified as bituminous.

Northern Appalachian Coal Field

The Northern Appalachian Coal Field, which encompasses Pennsylvania, northern West Virginia and Ohio is the most productive of the three Appalachian Fields. Within Pennsylvania, the second ranking coal producing state in 1959, there were 30 counties producing bituminous coal. However, production was concentrated in the 4 southwestern counties, Allegheny, Fayette, Washington and Greene. These 4 counties, having access to water transportation, produced 42 per cent of the Pennsylvania bituminous coal.[1] As shown in Table 10 and Figure 6, coal shipments in 1959, originating on waterways in Allegheny, Fayette, Greene and Washington Counties, accounted for 16,348,000 tons. Over 60 per cent of the coal production from these counties was marketed via the waterways. Most of this coal is high quality coking coal from the Pittsburgh Seam which is produced from captive mines owned by some of the major steel producers. Private and contract carriers transport most of this coal via water from the captive mines to coke plants located adjacent to navigable waterways.

West Virginia has been the leading coal producing state since 1931, and mines coal from both the Northern and Middle Appalachian Coal Fields. Monongalia and Marion Counties in northern West Virginia, which produced 7,277,000 and 9,348,000 tons, respectively, in 1959, form an important coal producing area. Over 4,538,000 tons (Table 10), or 62 per cent of the coal produced in Monongalia County, was marketed via the Monongahela River. Water transportation of coal from Marion County was unimportant, only 33,000 tons were shipped via water. In 1959, the upper Monongahela River which serves Marion County had a 7-foot channel which restricted the capacity of the coal barges loaded in this segment of the river.

The northern West Virginia Panhandle, which includes Marshall, Ohio and Brooke Counties, forms an important coal producing region. Coal production in 1959 in Marshall County was over 2,365,000 tons, in Ohio County, exceeded 1,199,000 tons, whereas, production in Brooke County was only 371,000 tons. Despite the location of these counties adjacent to the upper Ohio River, relatively small amounts of coals from this region are marketed via the waterways (Table 10). Most of the production in these counties is con-

[1] Bureau of Mines, *Mineral Yearbook*, 1959, Vol. 2, Fuels, (Washington, D.C., 1960), p. 110.

sumed locally in the Wheeling Industrial Node. Marshall County, the leading coal producing county in the West Virginia Panhandle, shipped 597,000 tons via the Ohio River in 1959. This was nearly 25 per cent of the county's total production.

In central West Virginia, Kanawha and Fayette Counties form another important coal producing area which is served by the Kanawha River. In 1959, coal production in these two counties amounted to 9,035,000 and 4,462,000 tons, respectively. In these counties most coal is produced by drift mining. However, in recent years auger mining has been increasing in importance, particularly in Kanawha County, where in 1959 production by auger mining was in excess of 745,000 tons.[2] Significant shipments of coal are marketed via the Kanawha River from Kanawha County, a total of 2,576,000 tons in 1959, or 28 per cent of its entire output. Over 1,611,000 tons or 36 per cent of the Fayette County coal output was shipped on the Kanawha River.

Ohio ranked fifth among the leading coal producing states, in 1959, production totaling 35,112,000 tons. The coal deposits are the most western continuation of the Northern Appalachian Coal Fields. Most of the coal seams are thinner than in the West Virginia area, but the Ohio coals are relatively near the surface so that strip mining is used in some areas. Eight counties in southeastern Ohio, which border the navigable Ohio River, produced slightly over 40 per cent of the State's total production. Belmont County, Ohio's second ranking coal producing county, which in 1959 produced 18.6 per cent of total Ohio coal production, transported 2,617,000 tons of coal, or 40 per cent of its production via the Ohio River (Table 10). Approximately 30 per cent of the Belmont County coal production is strip mined. Coal production in Jefferson and Columbiana Counties was 3,408,000 and 1,556,000 tons, respectively. In Columbiana County, which did not ship any coal via the Ohio River, nearly 90 per cent of production was recovered by strip mining. In Jefferson County only 7 per cent of production is marketed via the waterways, and 66 per cent of the recovery is by strip mining. Water transport is very important to Gallia County which shipped 97 per cent of

[2] Bureau of Mines, *Mineral Yearbook*, 1959, Vol. 2, Fuels, (Washington, D.C., 1960), p. 84.

TABLE 10—Bituminous Coal Shipped on the Ohio River System in 1959, by River and County of Origin
(In Net Tons)

State and County	River							Total
	Allegheny	Green	Kanawha	Kentucky	Monongahela	Ohio	Tennessee	Tradewater	
Illinois:									
Gallatin						131,531			131,531
Saline						95,872			95,872
Total						227,403			227,403
Indiana									
Warrick						2,050,215			2,050,215
Total						2,050,215			2,050,215
Kentucky:									
Daviess						556,581			556,581
Lee				24,000					24,000
Muhlenberg		4,339,854							4,339,854
Ohio		1,088,888							1,088,888
Perry				31,676					31,676
Union						2,644,207			2,721,938
Webster						488,174		77,731	488,174
Total		5,428,742		55,676		3,688,962		77,731	9,251,111
Ohio:									
Athens						57,038			57,038
Belmont						2,617,731			2,617,731
Gallia						830,894			830,894
Jackson						1,730			1,730
Jefferson						243,253			243,253
Meigs						319,079			319,079
Total						4,069,725			4,069,725

COAL PRODUCTION 25

TABLE 10—(Continued)

State and County	River							Total
	Allegheny	Green	Kanawha	Kentucky	Monongahela	Ohio	Tennessee	Tradewater	
Pennsylvania:									
Allegheny	1,431,651				145,433				1,577,084
Fayette					1,507,568				1,507,568
Greene					8,651,858				8,651,858
Washington					4,611,709				4,611,709
Total	1,431,651				14,916,568				16,348,219
Tennessee:									
Hamilton							9,871		9,871
Marion							524,832		524,832
Sequatchie							152,270		152,270
Total							686,973		686,973
West Virginia:									
Boone			467,573						467,573
Brooke						104,249			104,249
Fayette			1,611,476						1,611,476
Kanawha			2,576,916						2,576,916
Marion					33,595				33,595
Marshall						597,171			597,171
Mason						207,389			207,389
Monongalia					4,538,484				4,538,484
Ohio						9,061			9,061
Total			4,655,965		4,572,079	917,870			10,145,914
Total United States	1,431,651	5,428,742	4,655,965	55,676	19,488,647	10,954,175	686,973	77,731	42,779,560

Source: U. S. Bureau of Mines.

FIGURE 6—Shipmen of Bituminous Coal, by Counties Bordering Ohio River Waterways, 1959

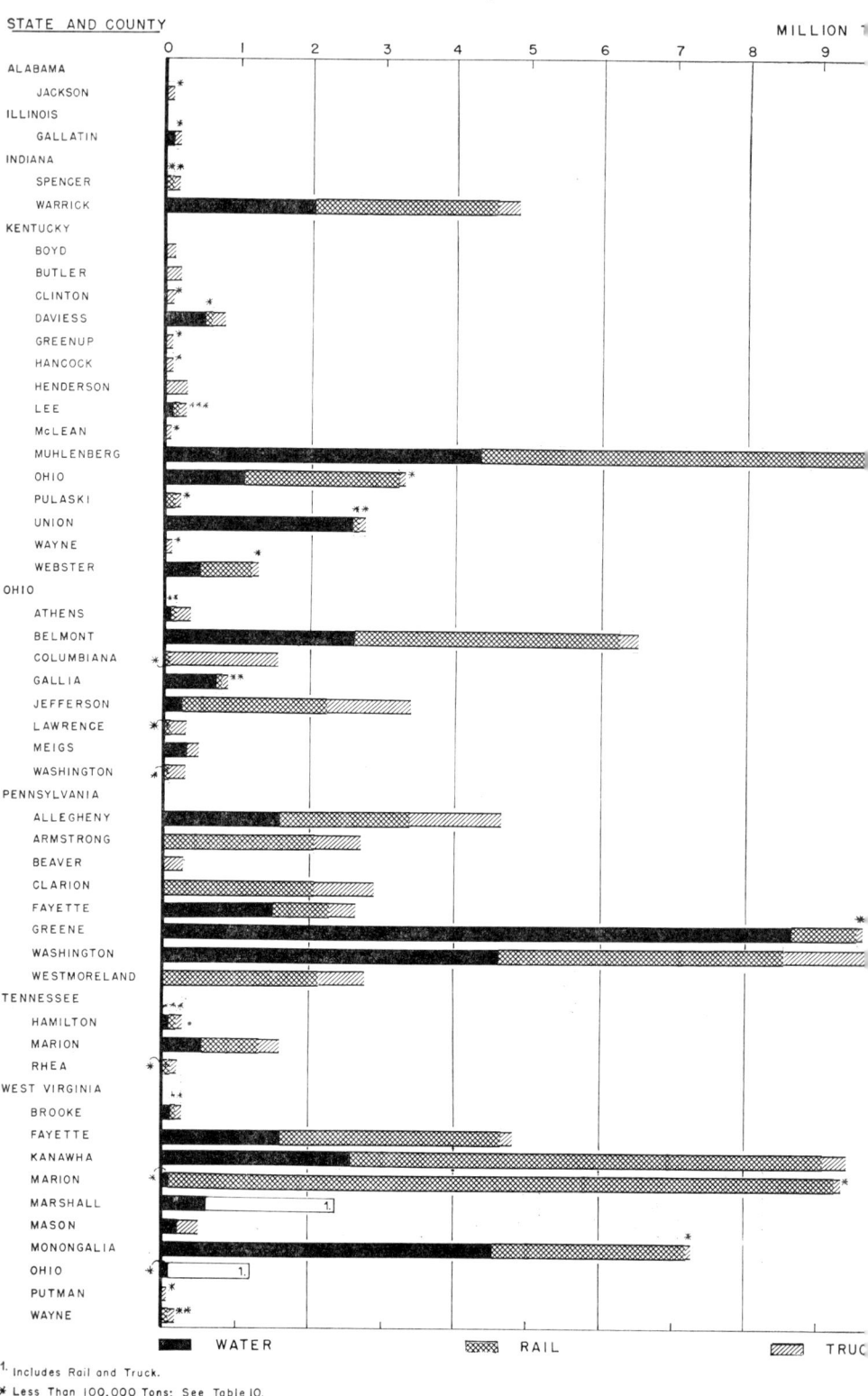

the county's coal production of 861,000 tons via the Ohio River. This is also an area where the bulk of the coal is strip mined. Other coal producing counties in Ohio which border the Ohio River include Athens, Lawrence, Meigs and Washington. Production of coal in 1959 in each of these counties was between 291,000 and 486,000 tons. Strip mining is an important method of mining in all of these counties but only in Meigs County is water transport of coal of any significance.

Eastern Interior Coal Field

Kentucky, Illinois and Indiana produce coal from the Eastern Interior Coal Field, the most productive field of the entire Interior Province. Although total coal production is considerably less than in the Appalachian Province, the Interior Coal Province is the second ranking coal producing area in the Western Hemisphere. There are several marked contrasts between the Eastern Interior and the Appalachian Coal Fields. The coal seams of the Eastern Interior Field are sufficiently near a more level surface in several areas to permit extensive strip mining. As a result of this type of production the coal output is extremely high in some places, averaging over 40 tons per man per day. In 1959, approximately 53 per cent of coal produced in this region was strip mined. However, the quality of these coals is generally somewhat inferior to the coal produced in the Appalachian Fields; at the present level of technology, most of these coals are not considered as coking quality.

Illinois is the leading coal producing state in the Eastern Interior Coal Field, ranking fourth in the Nation; the major regions of production, however, are not in the Ohio River Valley. Within Illinois only two counties currently ship coal via the Ohio River System. In 1959, Gallatin County shipped 131,531 tons via water. An unusual condition occurs in Saline County, Illinois, which is a relatively important coal producer with 1959 production of 2,570,000 tons. This county does not border a navigable waterway, but shipped 95,872 tons of coal via the Ohio River. Supplementary transport was used to move the coal across another county in order to utilize water transportation.

Only one county in the state of Indiana ships coal via the waterways. However, this county, Warrick County, is the State's leading producer of coal, producing in 1959 a total of 4,854,000 tons, of which 2,050,000 tons were marketed via the Ohio River. Most of the production in Warrick County is by strip mining which exceeded 4,468,000 tons in 1959. According to the Bureau of Mines, the coal production by strip mining in Warrick County was over 46 tons per man per day.

The western Kentucky segment of the Eastern Interior Coal Field is fortunate to be served by 4 navigable waterways, the Ohio, Green, Tradewater and Cumberland Rivers. *The Minerals Yearbook* 1959 of the Bureau of Mines lists 6 counties in western Kentucky each with annual coal production in excess of 800,000 tons, all 6 of which border navigable waterways. Five of these 6, namely, Daviess, Muhlenberg, Ohio Union and Webster Counties, utilize river transportation. In 1959, Muhlenberg County, on the Green River, was one of the two leading coal producing counties in the entire Eastern Interior Coal Field, with a production of 9,822,000 tons. Over 75 per cent or 7,458,000 tons of this coal was strip mined. Water transport was also important to Muhlenberg County; over 4,339,000 tons were shipped via the Green River. The other western Kentucky counties which are major coal producers and are served by water transport include Ohio, Union, Webster and Daviess. These counties produced 3,295,000, 2,735,000, 1,272,000 and 804,000 tons, respectively. With the exception of Union County, over 95 per cent of the coal was produced by strip mining. In Union County strip mining is negligible. Water transportation is of major importance to all of these counties, particularly Union County, which ships over 95 per cent of its coal production via the Ohio River.

Middle Appalachian Coal Field

Eastern Kentucky coal, which is located in the Middle Appalachian Field, does not contribute large quantities of coal to the Ohio River System traffic. Small amounts of coal from Perry and Lee Counties are transported on the Kentucky River. Production in Perry County in 1959 was 4,088,000 tons. It is surprising that any coal from Perry County is shipped via water since coal must be moved by supplemental transport across another county to reach a navigable waterway. Annual coal production in Lee County in 1959 was only 113,000 tons.

Southern Appalachian Coal Field

In Tennessee, the leading coal producing county, Marion County, is served by the Tennessee River. Marion County produced over 1,607,000 tons of coal in 1959. Most of the coal production is from the southeastern part of the State in the Southern Appalachian Coal Field. Approximately 12 per cent of this production was strip mined. About 33 per cent of the coal produced in Marion County was marketed via the Tennessee River. Other counties near the upper Tennessee River which are coal producers and ship via water include Hamilton and Sequatchie Counties.

Coal production in Alabama counties bordering the Tennessee River is insignificant. The only producer is Jackson County with a 1959 production of about 14,000 tons.

IV

GOVERNMENT REGULATION OF RIVER TRAFFIC

Coal transporters operating on the Ohio River System are classified into three categories: the private carrier, the contract carrier and the common carrier. Most coal is transported by private carriers and there are no special regulations on service or rates which apply to these carriers, since the transporter is the owner of the cargo, or the transporter is a wholly-owned subsidiary of a parent corporation which is the owner of the cargo. The contract carrier is next in importance according to the volume of coal shipped. As defined in the Interstate Commerce Act, "a contract carrier by water means any person (a carrier) which, under individual contracts or agreements, engages in the transportation by water of . . . property in interstate . . . commerce for compensation.[1] This type of carrier contracts to transport all or a specific amount of coal to a certain destination for a given period. Rates charged by a contract carrier are usually less than rates charged by a common carrier. Although a contract carrier may enter into an unlimited number of contracts, each particular tow is limited to only three commodities.[2] Almost all contract carriers operating on the Ohio River System transport but one commodity.

Finally, the common carrier, the least important in volume of the coal carriers on the Ohio River System, is the most regulated carrier. Rates and services, which are under the jurisdiction of the Interstate Commerce Commission, are published by the carrier. Common carrier rates are normally the highest of all carriers.

Other rules and regulations apply uniformly to all classes of carriers. The "Inland Rules of the Road" always apply on the Ohio River System with respect to piloting regulations. The United States Coast Guard is authorized to issue and rescind licenses to personnel operating vessels, and to inspect vessels. In case of damage by a vessel

---

[1] U. S. Congress, Senate, *The Interstate Commerce Act*, Sen. Doc. No. 72, 82 Cong., 1 Sess., Part III, Section 302 (e), p. 168.
[2] *Ibid*, Section 303 (b) p. 170.

to navigational installations it is the responsibility of the owner or leasee of the damaging vessel to make restitution.

RATES AND COSTS OF WATER SHIPMENTS OF COAL

As illustrated in Tables 11 and 12 and Figure 7 the primary reason for the increase in water transportation of coal has been the cost factor.

TABLE 11—River and Rail Rates per Ton for the Transportation of Coal between Selected Points[a]

	To:	From:	
	Cincinnati, Ohio	Huntington, W. Va.	Parkersburg, W. Va.	Powhatan Station, Ohio
River	$0.73	$1.30	$1.36
Rail	3.21[b]	3.21[b]	4.03

[a] In the tabulation of the rates for transportation of one ton of coal several factors must be specified. The effective date of these rates is Nov. 1, 1960. All quoted rates are common carrier rates. River rates are based on delivery to customer's terminal in barges with a 750 ton minimum. Rail rates are carload rates, loaded to 90 per cent of the marked car capacity (100,000 to 140,000 pounds) delivered to customer.

[b] Rail rates for coal from Huntington and Parkersburg are identical since both cities are located within the "Inner Cresent."

TABLE 12—Ton-Mile Common Carrier Rates for Coal and Distances for River and Rail Between Selected Points[a]

	To:	From:	
	Cincinnati, Ohio	Huntington, W. Va.	Parkersburg, W. Va.	Powhatan Station, Ohio
Distance via River		162 miles	286 miles	360 miles
River Rate		4.5 mills	4.5 mills	3.8 mills
Distance via Rail		162 miles	193 miles	308 miles
Rail Rate		19.8 mills	16.6 mills	13.1 mills

[a] Footnote *a* applies to tables 11 and 12.

Water transport costs by private carriers and rates charged by contract carriers are usually less than the rates charged by common carriers, hence, savings by use of water transport over rail for coal movements are actually greater than indicated.

Several conditions must be satisfied in order that a shipper may benefit from the lower transportation costs as offered by water shipments. First, the consumer must be in a position to handle large quantities such as transported in a river barge. As noted in Table 11 the minimum weight for a barge load rate is 750 tons. The consumer must be located so that the site has direct access to a river terminal. Unloading of a river barge usually requires a significant investment in equipment. If a major coal consumer is in a position to take

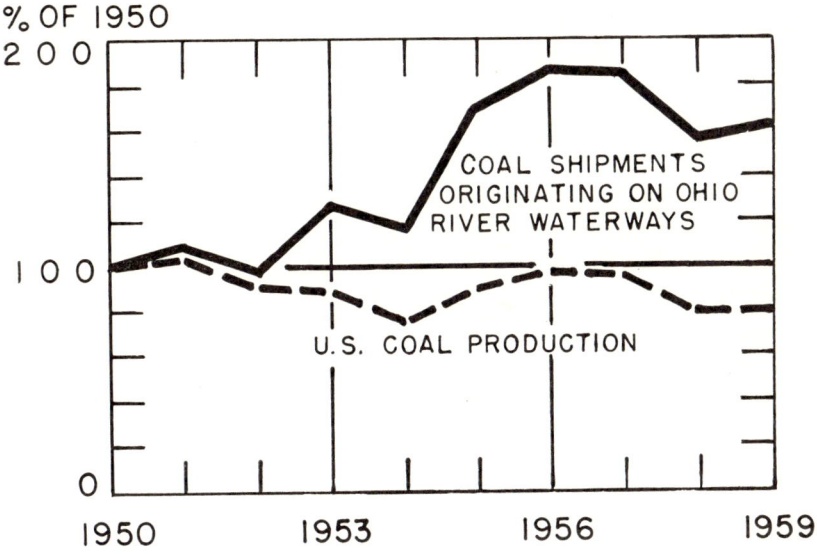

FIGURE 7—Relative Changes in Bituminous Coal Traffic on Ohio Waterways and U.S. Coal Production, 1950–1959

Source: Computations from data of U.S. Bureau of Mines.

advantage of water transportation, a marked reduction in transportation costs may be obtained.

With reference to Table 11, which is based on common carrier rates, it should be noted that the river rates in many instances are "paper rates," rates which are quoted in the existing river tariffs, but there may be little or no movement of this commodity via a common carrier. However, coal shipments can be made at the "paper rate," but usually private or contract carrier costs are less. The preceeding situation is not unusual with coal shipments on the Ohio River System where private and contract carriers transport the bulk of the coal.

An example of the cost of transporting coal via a private carrier for a distance of approximately 80 miles from the coal tipple to point of consumption on the Monongahela River was 4.3 mills per ton-mile in April 1961.[3] An immediate comparison with the rates charged by the common carriers as listed in Table 11, reveals a very slight difference between common carrier rates and private carrier costs. However, these data are not directly comparable. Normally, the costs per ton-mile decrease as the distance increases.[4] Where the ton-mile rates for common carriers decreases from 4.5 mills to 3.8 mills, the distance increases from 286 miles to 360 miles. The private carrier cost, 4.3 mills per ton-mile for coal, was for an 80 mile haul, whereas the common carrier rates are for greater distances. Since there is economy of scale in coal transportation as indicated by the trend towards larger barges and more powerful towboats, the cost of operations on the Ohio River are less than on the Monongahela River because of the size and number of barges in a tow. Due to the size of the lock chambers, channel depths and the meanders in the upper river, the standard Monongahela coal tow consist of 6 barges each measuring 175 by 26 feet and transporting 5,400 tons of coal. A typical Ohio River coal tow uses the "Jumbo Barge," which measures 195 by 35 feet, and the tow transports over twice as much coal, hence, costs per ton-mile are less on the Ohio River than on the Monongahela River. As a result of the short haul of 80 miles and limited size of the tows, the 4.3 mills per ton-mile costs for transporting coal can be considered somewhat above average costs for the Ohio River System.

Terminal costs for loading and unloading coal to and from river barges should be considered as these costs are directly related to the transportation costs. Correspondence with an executive of a private carrier on June 1, 1961 which also owns and operates several coal mines elicited comment on terminal charges as follows:

[3] This figure was made available by one of the major private carriers who preferred to remain anonymous, hence, the authors wish to honor this request.
[4] Meyer, J. R., Peck, M. J., Stenason, J., and Zwick, C., *The Economies of Competition in the Transportation Industries*, (Cambridge, 1959), p. 121.

Generally, you can figure on the dumping and loading of coal to run approximately 50 cents (a ton) and the unloading from the barge to a stockpile is about 27 cents (a ton). These costs that we use are for our own handling and generally prevail at some of the other terminals. Of course, it is hard to figure the outside costs because they are usually figured in the rail dumping rate or the price of the coal dumped into the barge.[5]

It should be stated that terminal costs for handling coal may be quite uniform with standardized equipment at most terminals on the Ohio River System. However, it must be assumed that the terminal equipment will be intensively utilized so that fixed expenses can be allocated over a large quantity of coal. The terminal costs as previously stated apply to terminals handling over 1,000,000 tons of coal annually.

[5] Since specific costs are stated, the source prefers to remain anonymous.

V

TRAFFIC FLOW

Coal traffic density on the Ohio River System varies considerably from one segment of a river to another as well as from one river to another. The maximum variation in the coal traffic density appears on the tributary waterways. On the Monongahela River this coal traffic attains the greatest density of all the rivers of the Ohio River System, whereas the navigable Cumberland River had no coal traffic in 1959, the date of the traffic-flow study.

As illustrated by Figure 8 all segments of the main-stem, the Ohio River from Pittsburgh to Cairo, had at least 2,000,000 tons of coal traffic. The coal traffic pattern on the Ohio River is relatively complex; in many instances coal shipments move in opposite directions in the same segment of the river. This pattern is a result of several factors. First, a substantial amount of the river transported coals are used to manufacture by-product coke for steel plants. The ideal coal for this process is a mixture of the high and low volatile bituminous coals; this necessitates shipping coal from various source areas with some shipments moving in opposite directions. A similar situation exists in some chemical plants. When coal is used as a basic raw material, the demand is for both the high and low volatile coals and for coals with other differences in chemical composition. Low-cost, strip-mined coal used in steam-electric plants may be shipped into an area which produces coking coal.

In an analysis of the coal traffic flow, it is essential to view the tributaries' coal shipments with the Ohio River coal traffic since the tributary traffic is integrated and directly related to the Ohio River traffic. All of the navigable tributaries, except the Cumberland and the Kentucky Rivers, serve as an origin or a destination for most of the Ohio River coal shipments (Figure 9). In 1959, the Ohio River tributaries were either a source or destination for 23,419,000 tons or 57.4 per cent of the Ohio River coal shipments. Coal shipments originating and destinted for Ohio River terminals in 1959 were 17,336,000 tons. The traffic flow on the Ohio River System is divided into three segments: the upper Ohio, above Powhatan Point; the middle Ohio, from Powhatan Point to Cincinnati; and the lower Ohio, from Cincinnati to Cairo. Since this division is based on the Ohio River System, the tributaries will be discussed with the appropriate subdivision.

UPPER OHIO RIVER

The upper Ohio River and its two important tributaries, the Monongahela and the Allegheny Rivers, have had a long and impressive history dating back to the late eighteenth and early nineteenth centuries as avenues of coal shipments. The maximum coal traffic density appears on the Monongahela River, where the major coal shipments are downstream. The traffic originates in the middle and upper Monongahela, below Morgantown, West Virginia. This *downbound traffic*[1] in 1959 was over 11,160,000 tons. The bulk of this coal is destined for the by-product coke ovens in the steel plants adjacent to the lower Monongahela River. Clairton, Pennsylvania with its huge coke ovens is the destination for much of this downbound coal traffic. In 1959, outbound coal traffic on the Monongahela was in excess of 8,675,000 tons. These shipments are classified as inbound traffic on the Ohio River and comprise 47.4 per cent of all inbound coal traffic on the Ohio River. Most of this coal is destined for the steel plants and steam-electric plants on the upper Ohio River, upstream from Steubenville, Ohio. In 1959, over 1,094,000 tons of coal moved inbound on the Monongahela River. Most of these shipments were low volatile bituminous coals originating on the Ohio and Kanawha Rivers. Nearly 1,454,000 tons of coal were shipped as outbound traffic from the lower Allegheny River to the markets on the upper Ohio River. Downbound coal shipments on the Allegheny River totaled 446,000 tons in 1959.

Coal traffic density on the upper Ohio River decreased downstream from a maximum of 11,000,000 tons at Pittsburgh to a mini-

[1] Inland waterway traffic terms pertaining to commodity shipments are defined as follows: downbound traffic, shipment terminates downstream on the same waterway as it originated; upbound traffic, shipment terminates upstream on the same water as it originated; inbound traffic, shipment originates on another waterway and terminates on specific waterway; outbound traffic, shipment originates on specific waterway and terminates on another waterway; and through traffic upbound or downbound, shipment moves upstream or downstream on the specific waterway but shipment originates and terminates on other waterways.

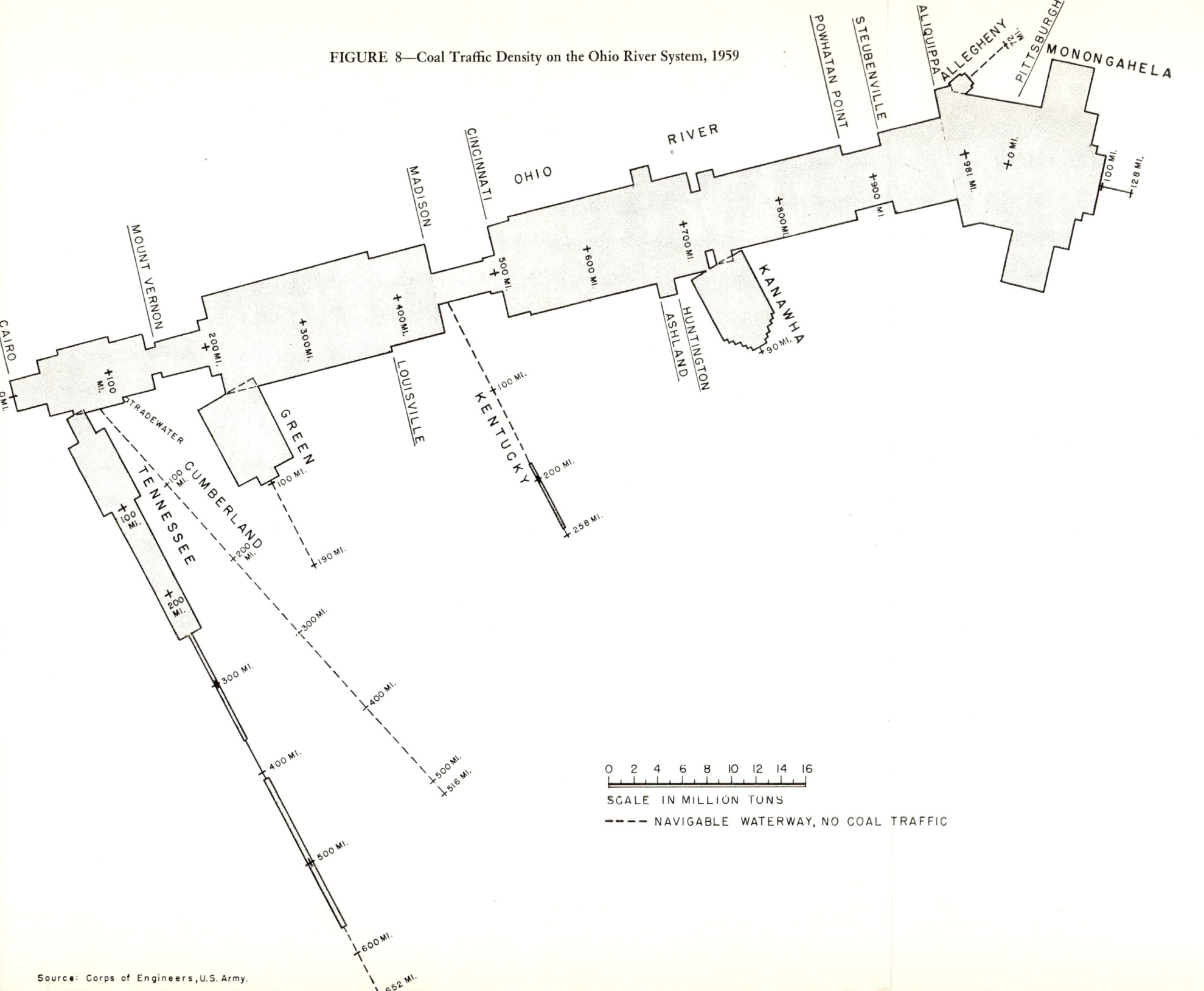
FIGURE 8—Coal Traffic Density on the Ohio River System, 1959

mum of 4,500,000 tons at Powhatan Point, Ohio. As previously mentioned this is a heavy industrial area which is basically a coal consuming region.

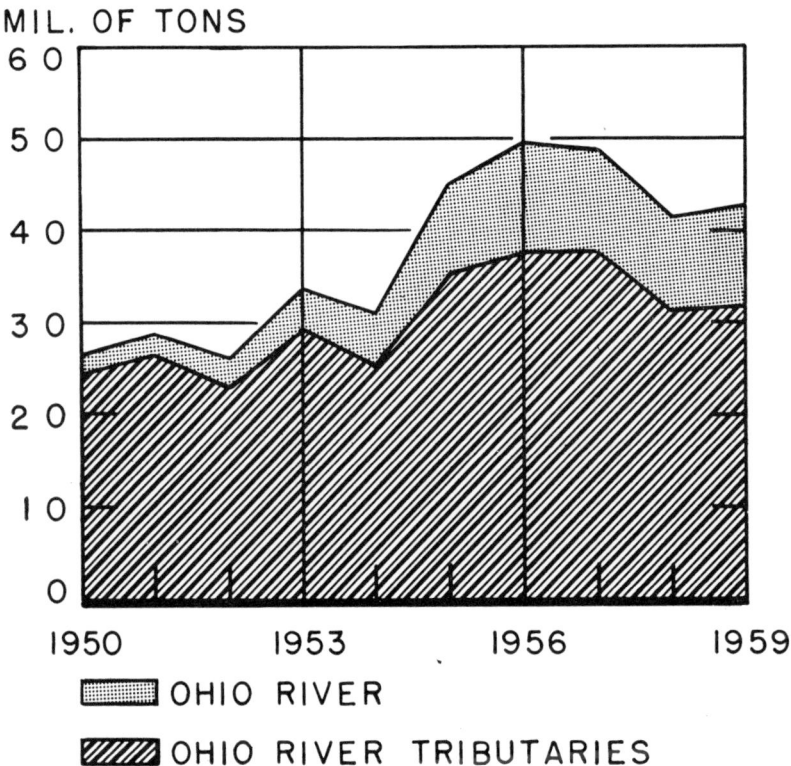

FIGURE 9—Origin of Coal Shipments on Ohio River System, 1950-1959

Source: Bureau of Mines.

MIDDLE OHIO RIVER

With reference to coal traffic on the middle Ohio River, from Powhatan Point to Cincinnati, the Kanawha River is the only tributary in this segment which contributes to the Ohio River coal traffic. The density of coal traffic in this segment of the Ohio River varies from a maximum of 10,500,000 tons downstream from Huntington to 4,800,000 tons immediately downstream of the mouth of the Kanawha River. Coal traffic density increases downstream from Powhatan Point as a result of coal shipments originating from Belmont County, Ohio. From this locale the coal traffic density is uniform for nearly 140 miles. The next change in the coal traffic density is caused by two coal consuming steam-electric plants, each having a capacity over one million kilowatts, being located upstream on the Ohio River from the mouth of the Kanawha River.

Coal shipments from the Kanawha River increase the upstream coal shipments on the Ohio River. On the Kanawha River, coal shipments are basically a downstream movement, with 87 per cent of the traffic classified as downbound or outbound shipments. Over 4,598,000 tons moved outbound for Ohio River markets; most of this coal was shipped upstream on the Ohio River. Downbound traffic on the Kanawha River was approximately 1,133,000 tons. Both the downbound and outbound coal traffic originate in Fayette and Kanawha Counties which are near the head of navigation on the Kanawha River. Combined upbound and inbound coal traffic on the Kanawha totaled less than 870,000 tons. The bulk of this movement, 805,000 tons, was upbound, destined for chemical plants and a steam-electric plant upstream from Charleston, West Virginia.

Coal traffic on the Ohio River increases a few miles downstream from the mouth of the Kanawha River and from this point to Cincinnati the traffic density varies moderately. The maximum density is attained in the segment of river between Huntington and Ashland. At Huntington, traffic increases downstream as coal shipped by rail from West Virginia coal fields enters the river traffic at a river-rail terminal. Downstream traffic decreases at Ashland, Kentucky, where the river transported coal is consumed by a chemical plant. From Ashland to Cincinnati, coal traffic is relatively consistent. Metropolitan Cincinnati receives a considerable amount of river-transported coal from upstream sources. Downstream from Cincinnati the traffic decreases to less than 3,000,000 tons. Steam-electric and chemical plants in the Cincinnati area are large consumers of river-transported coal. Some coal is transhipped at Cincinnati via rail to interior destinations in Ohio, such as the Miami Valley, Columbus and the Lake Erie ports.

LOWER OHIO RIVER

The lower segment of the Ohio River System extends from Cincinnati, Ohio to Cairo, Illinois, and includes 4 navigable tributaries, namely, the Tennessee, Green, Tradewater and Kentucky Rivers.

Coal traffic density on the lower Ohio River downstream from Cincinnati has great variations. In the segment of the river from Cincinnati to Madison, Indiana, the coal traffic is relatively light, averaging only 2,500,000 tons in 1959. However the traffic pattern changes at Madison, the site of one of two largest privately owned steam-electric plants in the world. This power plant depends on Ohio River transportation for its primary source of fuel. Much of this coal originates in the western Kentucky coal fields and enters the Ohio River as outbound traffic on the Green River. Coal traffic density is relatively consistent between Madison and the mouth of the Green River, a distance of approximately 230 miles. The only variation in the coal traffic is at Louisville which receives some of the upstream coal traffic for use in the metropolitan market which includes a steam-electric plant.

On the Green and Tradewater Rivers, the coal traffic pattern is relatively simple. In 1959, over 5,141,000 tons of coal was transported on the Green River; all of this traffic was downstream and over 96.5 per cent was classified as outbound and destined for steam-electric plants on the lower Ohio River. Muhlenburg and Ohio Counties in western Kentucky serve as the major sources of the Green River coal traffic. Approximately 177,000 tons of coal was shipped downbound on the Green River to a steam-electric plant. Union County, Kentucky shipped 101,000 tons of coal outbound from the lower Tradewater River into the Ohio traffic in 1959.

In the segment of the Ohio River below the mouth of the Green River, the coal traffic density shows great variations. This results from several activities. Coal traffic density decreases at Mount Vernon, Indiana which is the site of a river-rail terminal that transfers coal from barges to rail cars. Between Mount Vernon and the mouth of the Tennessee River, coal traffic increases as there are several mines in the Kentucky and Illinois counties which border the Ohio

River that market their coal via water transport. Coal traffic decreases at the mouth of the Tennessee River as the basic movement from the Ohio River is inbound on the Tennessee.

Total coal traffic on the Tennessee River in 1959 was 5,169,000 tons. This is comparable with the Green River, but the traffic pattern on the Tennessee is far more complex with coal shipments moving inbound, outbound, upbound and downbound. Due to its length the Tennessee River serves several areas with different demands and the average length of coal haul on the Tennessee is considerably longer than on the Green River. Over 2,965,000 tons, or slightly over 57 per cent of the total coal traffic on the Tennessee River, is classified as upbound. The bulk of this traffic originates at a river-rail terminal on the lower Tennessee. Inbound coal traffic can also be considered as a relatively important phase of the coal movement. In 1959, inbound coal shipments were 1,355,000 tons which comprised over 26 per cent of the total coal traffic. Outbound coal shipments in 1959 were 11.5 per cent of the total coal movements on the Tennessee River. The T.V.A. steam-electric plants at Johnsonville, Tennessee, 100 miles upstream from the mouth of the Tennessee, and the T.V.A. steam-electric plant at Pride, Alabama, 245 miles upstream, are the two major consumers of river transported coal. Other major coal terminating sites on the Tennessee River are Guntersville, Alabama, a transhipment center, and a T.V.A. steam-electric facility at Stevenson, Alabama.

The final decrease in the coal traffic density immediately upstream on the Ohio River from Cairo, Illinois, results from coal consumption at two large steam-electric plants. These plants located at Shawnee, Kentucky and Joppa, Illinois, almost across the river from one another, each have a listed capacity in excess of 1,000,000 kilowatts.

VI

CONSUMPTION OF BITUMINOUS COAL

The use of bituminous coal by type of consumer in the United States is shown in Figure 10.[1] A decrease in the consumption of coal by railroads and a marked increase in the use of coal by the electric utilities are the pronounced changes in types of consumers. In 1950,

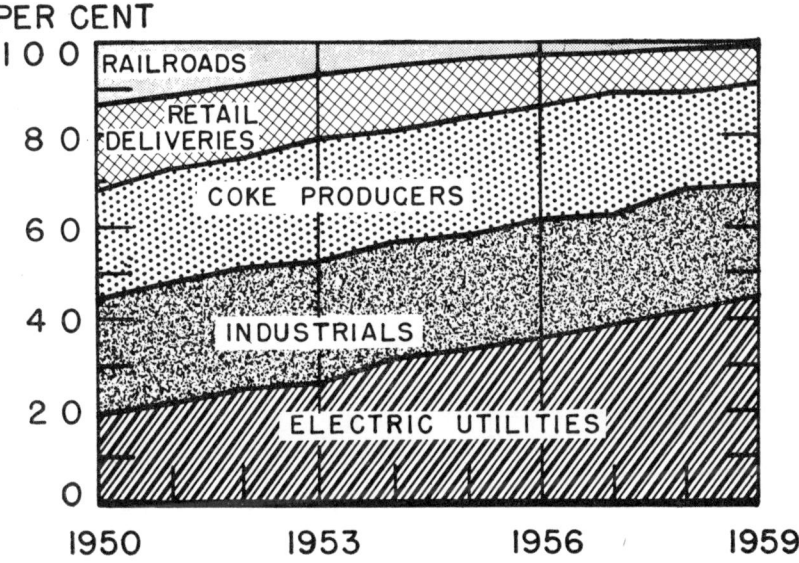

FIGURE 10—Percentage of Total Consumption of Bituminous Coal, by Type of Consumer in the United States, 1950–1959

Source: Bureau of Mines.

the railroads consumed 60,969,000 tons of coal as compared with only 2,600,000 tons in 1959, whereas the electric power utilities increased their consumption of coal from 88,262,000 tons in 1950 to 165,788,000 tons in 1959. The amount of coal used by the industrials over the past decade has fluctuated from 122,955,000 tons to

[1] Includes lignite.

88,580,000 tons while the coke producer's consumption varied from 113,448,000 to 76,580,000 tons. The retail delivery of coal for the 10-year period has dropped from 84,422,000 tons to 29,138,000 tons (Table 13).

Despite the growing efficiency of coal utilization and the future possibilities of nuclear power, a continued expanding demand for coal appears probable. In 1920 it required three pounds of coal per kilowatt-hour, whereas in 1959 less than one pound of coal was required to produce a kilowatt-hour (Figure 11). Although it takes

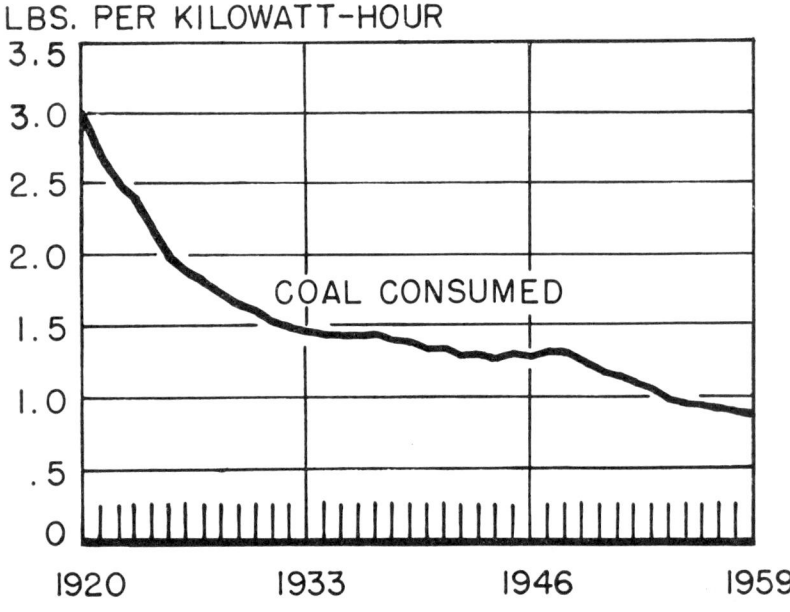

FIGURE 11—Consumption of Coal at Electrical-Utility Power Plants in the United States, 1920–1959

Source: Bureau of Mines.

less coal to produce a unit of electricity, the lowered unit cost resulting therefrom should enable the industry to sell more electricity with an over-all increase in the consumption of coal (Table 14). In fact, the Federal Power Commission has stated that by 1970 the

TABLE 13—Major Consumers of Coal in the United States
(In Thousands of Tons)

Year	Electric Utilities		Industrials		Coke Producers		Retail Delivery		Railroads		Total Consumption[a]
	Amount Consumed	Per Cent of Total	Amount Consumed	Per Cent of Total	Amount Consumed	Per Cent of Total	Amount Consumed	Per Cent of Total	Amount Consumed	Per Cent of Total	
1950	88,262	19.44	114,662	25.24	103,845	22.86	84,422	18.59	60,969	13.42	454,202
1951	101,898	21.73	122,955	26.22	113,448	24.19	74,378	15.86	54,005	11.52	468,904
1952	103,509	24.67	111,172	26.55	97,614	23.31	66,861	15.97	37,962	9.07	418,757
1953	112,283	26.31	112,091	26.26	112,874	26.45	59,976	14.05	27,735	6.50	426,798
1954	115,235	31.74	92,022	25.35	85,391	23.52	51,798	14.27	17,370	4.78	363,060
1955	140,650	33.19	105,493	24.91	107,377	25.36	53,020	12.52	15,473	3.65	423,412
1956	154,983	35.80	109,517	25.30	105,913	24.47	48,667	11.24	12,308	2.84	432,858
1957	157,398	38.05	102,773	24.84	108,020	26.11	35,712	8.63	8,401	2.03	413,668
1958	152,928	41.70	96,896	26.42	76,580	20.88	35,619	9.71	3,725	1.02	366,703
1959	165,788	45.26	88,580	24.19	79,181	21.62	29,138	7.96	2,600	.71	366,256

Source: Bureau of Mines.
[a] Total consumption includes bunker coal for foreign and lake vessels.

electric utility industry in the United States will need to expand its generating capacity over 1960 by 50 per cent to meet the increasing demands for electric power.

The Ohio Valley illustrates the nature of the expanding electric utilities and the increasing outlet for coal. There are 31 electric power stations along the Ohio River with a total generating capacity in excess of 12,500,000 kilowatts. Of the 31 power stations, 29 are steam-electric plants. One of the major power units is the Clifty Creek plant at Madison, Indiana with a capacity of 1,290,000 kilowatts, which, in 1959, consumed 4,225,000 tons of coal, all water

TABLE 14—Fuel Economy in Consumption of Coal at Electric-Utility Power Plants

Year	Coal Consumed Per Kilowatt-Hour (Pounds)	Index (1919=100)
1919	3.20	100.0
1920	3.00	93.8
1925	2.00	62.5
1930	1.60	50.0
1935	1.44	45.0
1940	1.34	41.9
1945	1.30	40.6
1950	1.19	37.2
1955	.95	29.7
1956	.94	29.4
1957	.93	29.1
1958	.90	28.1
1959	.89	27.8

Source: Bureau of Mines.

transported. Among the other plants of significance is the Kyger steam-electric plant at Cheshire, Ohio with a capacity of 1,070,000 kilowatts, which, in 1959, consumed 2,770,000 tons of coal, all shipped in barges (Figure 8).

In addition to the power plants on the Ohio River there are a number of steam-electric plants on the major tributaries of the Ohio. A report by the Tennessee Valley Authority states that 15 years ago the region's power needs could be supplied almost entirely from hydrogeneration. Now the demand for electricity has increased so that only about one-fourth of the needs are supplied by the hydroelectric plants; the rest is generated at steam plants. Coal is now the largest item in T.V.A. power production costs.

With respect to the relation of electric energy and nuclear power, it has been estimated that by the year 2000 only about 20 per cent of the total energy requirements of our Nation will be supplied through atomic energy, and that coal will provide 30 per cent of the total.

VII

SUMMARY

With the completion of the proposed navigational improvements and the continued development of the resources of the Ohio River Valley, the future industrial development of this area appears assured. However, if the waterways are to remain competitive with other modes of transport, the latest technological advancements must be incorporated in the improvement and maintanance of the navigational facilities on the Ohio River and its tributaries. At present the tributaries serve an important role as a source and/or destination for much of the Ohio River coal traffic. New and larger river dams which will increase length of pools and reduce the number of lockages and increase the speed of the traffic, the enlargement of the lock chambers which can accommodate larger tows and the increase of channel depths to facilitate the movement of larger vessels will aid in the development of a more efficient and competitive transportation on the Ohio River System.

Despite the fact that coal mining has declined in recent years, the transportation of coal on the Ohio River System has continued to increase (Figure 7). Coal reserves in the states adjacent to the navigable waterways of the Ohio River System are sufficient to meet the Nation's demands for many years. Although there are chemical differences in the coal which has access to water transportation, all of this coal is classified as bituminous and has from medium to high BTU content. The chemical differences in the coals are an advantage since coal can be found to fulfill many purposes. Some of the coal deposits are sufficiently near the surface to permit strip mining, and many of the other coal beds have characteristics which permit mechanization in underground mining.

At present, coal is the most important commodity (based on tonnage) in the traffic on the Ohio River System, and the demand for coal for generating electrical power, and for the manufacturing of steel and aluminum should continue to increase with the industrial development in the valleys of the Ohio River System.

3568 L.